ALPHAMETICS EXPRESSING THOUGHTS FROM THE STAR TREK ORIGINAL SERIES

Written by

Charles Ashbacher

5530 Kacena Ave

Marion, IA 52302

cashbacher@yahoo.com

Cartoon artwork

Caytie Ribble

Image of Spock

Jenna Richardson

ISBN-13: 978-1512152784

CONTENTS

Introduction

This book is one of many consequences of my love for the Star Trek™ original series. It was a ground-breaking show in many respects, the most discussed is almost certainly the supposed first inter-racial kiss between a white person and a black person that appeared on network television. That is not in fact the case, but it is certainly the most celebrated. The kiss between Captain Kirk and Lieutenant Uhuru took place in the third season episode "Plato's Stepchildren" and that show was also controversial over the depiction of sadism.

There is one alphametic for each of the 79 episodes of the original series plus the original pilot "The Cage." The messages of the alphametics are related to the title and plot of the episodes and are to be solved in base 10 unless otherwise noted.

Three cartoons with a Star Trek theme that were drawn by artist Caytie Ribble are also included. They, along with many other cartoons with a mathematical theme will appear in a future book. Explanations of the cartoons are included.

An image of Spock drawn by artist Jenna Richardson is included as a tribute to the late Leonard Nimoy.

I hope you enjoy these problems as much as I enjoyed creating them! Of course, Star Trek is a registered trademark of CBS Studios.

Charles Ashbacher
cashbacher@yahoo.com

Leonard Nimoy, 1931 - 2015

Cartoons
Drawn by Caytie Ribble

LANDRU ALGEBRA

$$A \cup (A \cap B) = A$$

$$A \cap (A \cup B) = A$$

Introduction to Alphametics

A **cryptarithm** is an arithmetic computation where the digits are replaced by letters. For example, the simple sum

$1234 + 5678 = 6912$

can be expressed in the cryptarithmic form

abcd + efgh = fiab

 When decoding cryptarithms, the standard rules are as follows:

1) The numbers are in base ten unless stated otherwise.

2) No leading digit can be zero.

3) Each letter represents a unique digit.

4) The number of solutions is small or there is an additional qualifier that makes the number small.

5) The operations are addition unless otherwise noted.

 Although the advent of computers has changed the soluton strategy somewhat, the general approach to solving cryptarithms is by using basic arithmetic and a bit of logic.

 An **alphametic** is a cryptarithm where the substitution yields an understandable message. For example, consider the following problem

```
              WE
             ALL
            NEED
            MORE
        _____

           MONEY
```

where of course we want the most MONEY.

Note: One condition that occurs quite often in alphametics is the possibility of interchanging digits. Since W and R are in the same column and appear nowhere else in the problem, if they are both nonzero their values can be interchanged without altering the result. Therefore, it is possible to have two "unique" solutions.

When searching for solutions, the place to start is often the position of the leading digits. In this case, the sum of N and M with the value carried in must be greater than 10 with a leading digit equal to M. Therefore, the only possible value for M is 1. The sum of the next position over

$$\text{Carry in} + A + E + O$$

could have at most a carry out of two and with N having the maximum value of 9, O would have a maximum value of 2. Since 1 is already used, O would have a value of 0 or 2. Also, this forces the value of N to be al least 7. If the carry out from this sum is 2, the sum would have to be at least 27, which is impossible. Therefore, the carry out from the sum

$$\text{Carry in} + A + E + O$$

is at most 1. This means that N must be at least 8 and O must be zero.

Given that W, L, E and R are all at least 2 and have different values, the carry out from the sum

$$\text{Carry in} + W + L + E + R$$

must be at least 1 and can be at most 2. With O equal to zero, the sum

$$\text{Carry in} + A + E$$

would have to sum to at least 18 to have a carry out. Since 8 or 9 is taken by N, the maximum value of $A + E$ is 16. Therefore, if N = 9, the sum must be 9 and if N = 8, the sum can be 18, but only if A and E are 9 and 7 and the carry in is 2. If the column sum is 8, then the values of A and E are restricted to smaller values.

It is here where we use the last bit of information about wanting the most MONEY. We would set N to 9 and proceed from there. It is most likely that the maximum of MONEY would be when N is 9.

Filling in all values currently known or surmised, we would have

```
            #   #
        #   #   #
    9   #   #   #
    1   0   #   #
   _____

    1   0   9   #   #
```

We also would expect the largest value to occur when the carry out from the 10's column is the largest it can be, which is 2. Therefore, we start with the case

Case 1 Assume N = 9 and the carry out from the 10's column is 2. This gives the sum

$2 + A + E = 9$

or

$A + E = 7$.

This yields two possible combinations for A and E, 2 and 5 or 3 and 4. In keeping with our goal to find the maximum, we would want E to be 5, which leads to the assignments of

$A = 2$ $E = 5$.

Filling in these numbers, we have

```
      #  5
   2  #  #
9  5  5  #
1  0  #  5
_____

1  0  9  5  #
```

Our next step will be to examine the rightmost column. The goal is to find the maximum value for MONEY, so we want the largest possible digit in Y. However, with 9 taken, the largest possible value for $L + D$ is 15, which is impossible, since 5 has been used. Therefore, the largest remaining choice for Y is 4. The only choices for this sum are 8 and 6. Since L also appears in the 10's column, the best choice is to assign L the digit 8. This leads to the sum

```
      #  5
   2  8  8
9  5  5  6
1  0  #  5
_____

1  0  9  5  4
```

which leaves the digits 7 and 3 for W and R.

```
      7 5
    2 8 8
  9 5 5 6
  1 0 3 5
  _____

  1 0 9 5 4
```

By virtue of our choice of values, this would be the largest possible value for the sum. It should be noted that other possible lines of reasoning will also lead you to this solution.

Gene Roddenberry, the creator of Star Trek mentioned in his book **The Making of Star Trek** that Captain James Tiberius Kirk was born in Iowa. In 1985, the small Iowa town of Riverside, with the approval of Roddenberry, proclaimed itself the future birthplace of James T. Kirk. This was the background for the first Star Trek alphametic that I created.

```
KIRK + BORN + HERE + IN = IOWA
```

where IOWA is maximal.

This problem was published in the alphametics section of **Journal of Recreational Mathematics**.

Limerick

by Charles Ashbacher and Jen Corrigan

There was a bold captain named Kirk

Who was often seen wearing a smirk

They embarked on a mission

With pride and ambition

The ladies were just a great perk

THE STAR TREK ALPHAMETICS

Each of the alphametics in this collection is based on an episode of the Star Trek Original Series. The first number is the episode number and that is based on the order of first appearance. Unless otherwise noted, the explicit digits in the problem can be reused.

We start with a tribute to the first pilot, "The Cage."

```
PILOT
  THE
 CAGE         the solution is to be in base 12 where CAGE
 PIKE         PIKE and PILOT are all prime
 ISAT
───────

TALOS
```

Tribute to episode 1, "The Man Trap."

```
    1
 SALT
   IS         where TRAP is the greatest
 AMAN
───────

 TRAP
```

Tribute to episode 2, "Charlie X"

```
      2
THASIAN
 POWERS      since the powers are strong, we maximize CHARLIE in
 WITHIN      this base 14 alphametic
───────

CHARLIE
```

Tribute to episode 3, "Where No Man Has Gone Before"

```
     3
 WHERE
 NOMAN       solve it in base 12 where nothing comes
   HAS       before BEFORE
  GONE
───────

BEFORE
```

Tribute to episode 4, "The Naked Time"

```
      4
  TIME
    NO    since deep personality traits are exposed we maximize NAKED
  DATE
  ──────
  NAKED
```

Tribute to episode 5, "The Enemy Within"

```
       5
   IN
   ONE              where KIRK is of course the greatest
   TWO
   OF
   ──────
   KIRK
```

Tribute to episode 6, "Mudd's Women"

```
      6
 (LEO)H    where MUDD is the lowest and the punctuation characters are not part
 MUDD'S    of the problem
 ──────────
   WOMEN
```

Tribute to episode 7, "What Are Little Girls Made Of?"

```
        7
    OF
   WHAT
    ARE    solve in base 16 and make GIRLS as great as possible and
   THEY    MADE is prime
   MADE
   ──────
   GIRLS
```

Tribute to episode 8, "Miri"

```
      8
  MIRI
   HAS    since a girl CRUSH can be so hard, we make it the largest
     A
  KIRK
  ──────
  CRUSH
```

Tribute to episode 9, "Dagger of the Mind"

```
            9
      FIRST
       MIND      solve in base 12 and for the largest DAGGER
      MELDA
       MIND
      _____

      DAGGER
```

Tribute to episode 10, "The Corbomite Maneuver"

```
      10
   POKER
     NOT  since POKER is an "interesting" game we maximize it
   CHESS
   _____

   BALOK
```

Tribute to episode 11, "The Menagerie Part I"

```
          11
      SPOCK
       PIKE    of course, SPOCK is the greatest
       BACK
         TO
      _____

      TALOS
```

Tribute to episode 12, "The Menagerie Part II"

```
      12
   PIKE
   BACK      where PIKE is greatest
   WITH
   _____

   VINA
```

Tribute to episode 13, "The Conscience of the King"

```
      13
     HOW
    DOES      we will of course minimize KODOS
      HE
    LIVE
    _____

   KODOS
```

Tribute to episode 14, "Balance of Terror"

```
      14
   TERROR
  ISNOTIN     solve in base 11 and maximize the TERROR
  _____

  BALANCE
```

Tribute to episode 15, "Shore Leave"

```
      15
    CREW
    HASA      since all want to spend as long as they can on  leave, we want
   SHORE      to maximize LEAVE
   _____

   LEAVE
```

Tribute to episode 16, "The Galileo Seven"

```
      16
   GANGOF
  SEVENON
  _____

  GALILEO
```

Tribute to episode 17, "The Squire of Gothos"

```
      17
     THE
   SQUIRE     solve in base 14 and since Trelane is not a real
      OF      squire, minimize SQUIRE
   GOTHOS
   _____

  TRELANE
```

Tribute to episode 18, "Arena"

```
        18
      KIRK    where KIRK is the greatest
      GORN
        IN
      _____

     ARENA
```

Tribute to episode 19, "Tomorrow is Yesterday"

```
          19
     ATOMORROW
     THETIMEIS
     _____

     YESTERDAY
```

Tribute to episode 20, "Court Martial"

```
        20
        HE    solve in base 12 and of course,
      WINS    SPOCK is the greatest
        AT
     CHESS
     _____

     SPOCK
```

Tribute to episode 21, "The Return of the Archons"

```
       21
     LANDRU
        DID     solve in base 12
     ABSORB
     _____

     ARCHONS
```

Tribute to episode 22, "Space Seed"

```
       22
     KHAN
       ISA   we acknowledge the vastness of SPACE by making it maximum
     SEED
       IN
     _____

     SPACE
```

Tribute to episode 23, "A Taste of Armageddon"

```
      23
    AWAR
    WITH     solve in base 14 and minimize STOPS
      NO
     END
    KIRK
   ───────
   STOPS
```

Tribute to episode 24, "This Side of Paradise"

```
      24
     THE
    KISS     where SPOCK is of course the greatest
    THIS
    SIDE
   ───────
   SPOCK
```

Tribute to episode 25, "The Devil in the Dark"

```
      25
     MOM
   HORTA      solve in base 14 and minimize DEVIL
   ISNOT
       A
   ───────
   DEVIL
```

Tribute to episode 26, "Errand of Mercy"

```
       26
       NO
    MERCY     solve in base 12 and maximize ORGANIA
   ERRAND
   NEEDED
       ON
   ────────
   ORGANIA
```

Tribute to episode 27, "The Alternative Factor"

```
            27
         THERE
        ARETWO     solve in base 16 and maximize LAZARUS
        OF THEM
        ONEGOOD
        _____

        LAZARUS
```

Tribute to episode 28, "The City on the Edge of Forever"

```
           28
         KIRK
          SHE    of course, KIRK is the greatest
        HASTO
          DIE
        _____

        EDITH
```

Tribute to episode 29, "Operation Annihilate!"

```
         29
       SPOCK
          IS   solve in base 14 and make SPOCK the greatest
       BLIND
          ON
       _____

     DENEVA
```

Tribute to episode 30, "Amok Time"

```
            30
           PON
          FARR   solve in base 12 and maximize MATE
           DIE
            IF
           NOT
          _____

          MATE
```

Tribute to episode 31, "Who Mourns for Adonais"

```
          31
HOWCAN    make APOLLO the greatest, odd god
   HEBE
THEGOD
_____

   APOLLO
```

Tribute to episode 32, "The Changeling"

```
        32
  JR
KIRK
  JT
KIRK
_____

  NOMAD
```

Tribute to episode 33, "Mirror, Mirror"

```
          33
  STAR
   TREK
 MIRROR    where TREK is the greatest!
 MIRROR
_____

   EMPIRE
```

Tribute to episode 34, "The Apple"

```
      34
 EYES
 EARS    where we want the greatest APPLE
   OF
 VAAL
_____

 APPLE
```

Tribute to episode 35, "The Doomsday Machine"

```
        35
       THE
      BOMB   solve in base 12 and maximize BOMB
      USED
       FOR
      SOME
     _____

      GOOD
```

Tribute to episode 36, "Catspaw"

```
         36
      SYLVIA
       KOROB    solve in base 14 and maximize the evil SYLVIA
       PLAYA
     _____

     CATSPAW
```

Tribute to episode 37, "I, Mudd"

```
        37
       HIS
      NAME      Let's humor MUDD by making him the
       ISI      greatest
     _____

      MUDD
```

Tribute to episode 38, "Metamorphosis"

```
        38
      ALIEN
     ENTITY  solve in base 14 and maximize ZEFRAM
     AYOUNG
     _____

     ZEFRAM
```

Tribute to episode 39, "Journey to Babel"

```
        39
      SAREK
        AND  solve in base 12 and maximize the size of Spock's PARENTS
     AMANDA
     SPOCKS
     _____

     PARENTS
```

Tribute to episode 40, "Friday's Child"

```
        40
      NEW
     TEER    solve in base 12 and since the child will be Teer
    CHILD    maximize CHILD
       OF
    _____

    ELEEN
```

Tribute to episode 41, "The Deadly Years"

```
       41
    AGED
     AND    naturally we will maximize FAST
     DIE
      SO
    _____

    FAST
```

Tribute to episode 42, "Obsession"

```
       42
     ALL
    BLOOD    here we will maximize the value of KILLER
    EATER
      ALL
    _____

    KILLER
```

Tribute to episode 43, "Wolf In the Fold"

```
       43
     WOLF
      IN    here we maximize FOLD
     THE
    _____

    FOLD
```

Tribute to episode 44, "The Trouble With Tribbles"

```
        44
      KIRK
       AND        solve in base 12 and maximize KOLOTH
     LURRY
      ATK7
   ANDADD
   ————————

   KOLOTH
```

Tribute to episode 45, "The Gamesters of Triskelion"

```
        45
      KIRK
     UHURA     solve in base 12 and maximize SLAVES
    CHEKOV
       ARE
   ————————

   SLAVES
```

Tribute to episode 46, "A Piece of the Action"

```
         46
          A
      PIECE
      OFTHE
   ————————

   ACTION
```

Tribute to episode 47, "The Immunity Syndrome"

```
        47
        AN
    ENERGY     solve in base 12 and maximize AMOEBA
     DRAIN
     LARGE
   ————————

   AMOEBA
```

Tribute to episode 48, "A Private Little War"

```
      48
ALITTLE    solve in base 12 and since it will not remain
WARITIS    private, we will minimize PRIVATE
_____

PRIVATE
```

Tribute to episode 49, "Return to Tomorrow"

```
      49
       A
     WISE    solve in base 12 and maximize SARGON
  ANDOLD
   ALIEN
    MIND
  _____

  SARGON
```

Tribute to episode 50, "Patterns of Force"

```
    50
    ON
  EKOS    we will maximize MESS here
  ZEON
  GILL
  _____

  MESS
```

Tribute to episode 51, "By Any Other Name"

```
   51
  KIRK    solve in base 14
  MUST    and maximize the value of FIGHT
 FIGHT
 _____

 ROJAN
```

Tribute to episode 52 "The Omega Glory"

```
    52
   USA
  USSR    given the weakness of the episode, GLORY is minimal
 OMEGA
 _____

 GLORY
```

Tribute to episode 53, "The Ultimate Computer"

```
      53
 M5KIRK    solve in base 12 and minimize DUNSAIL
 ISNOTA
 _____

 DUNSAIL
```

Tribute to episode 54, "Bread and Circuses"

```
      54
 ROME
    IS  we of course maximize ROME
 ONCE
 _____

 AGAIN
```

Tribute to episode 55, "Assignment Earth"

```
      55
 ISIS
   AND    solve in base 12 and maximize the value of EARTH
 GARY
 SEVEN
 _____

 EARTH
```

Tribute to episode 56, "Spock's Brain"

```
      56
 NO
 BRAININ    solve in base 12 and since this is likely the worst episode
  SPOCK     of the original series, minimize EPISODE
 NO
 BRAININ
 _____

 EPISODE
```

Tribute to episode 57, "The Enterprise Incident"

```
   57
 KIRK
 MUST    solve in base 14 and maximize the value of the CLOAK
GETTO
  ROM
-------

CLOAK
```

Tribute to episode 58, "The Paradise Syndrome"

```
   58
 KIRK
TAKES    since KIROK is not a real character, the value is minimal
AWIFE
   AS
-------

KIROK
```

Tribute to episode 59, "And the Children Shall Lead"

```
   59
SCARY
FORCE
 FEAR
--------

GORGAN
```

Tribute to episode 60, "Is There in Truth No Beauty?

```
    60
KOLLOS    solve in base 14 and maximize INSANE
DRIVES
 SPOCK
-------

INSANE
```

Tribute to episode 61, "Spectre of the Gun"

```
      61
   EARP
   EARP   solve in base 14 and maximize the FIGHT
   EARP
    DOC
   AGUN
  _____

   FIGHT
```

Tribute to episode 62 of the Star Trek original series, "Day of the Dove"

```
     62
  DAY
   OF      where the peace makes it the greatest DAY
  THE
  _____

  DOVE
```

Tribute to episode 63, "For the World is Hollow and I Have Touched the Sky"

```
       63
   NATIRA
      AND   given the size of the hollow asteroid we maximize
    MCCOY   YONADA
       ON
  _____

   YONADA
```

Tribute to episode 64, "The Tholian web"

```
      64
   KIRK
  ISOUT   solve in base 12 and maximize the PHASE
     OF
  _____

  PHASE
```

Tribute to episode 65, "Plato's Stepchildren"

```
     65
  KIRK
 UHURU    while it was not the first B & W kiss on television, it was one of the
  KISS    first, so minimize FIRST
   NOT
 _____

 FIRST
```

Tribute to episode 66, "Wink of an Eye

```
      66
   KIRK
     TO    we maximize the SPEED
  SPEED
     TO
  _____

  BREED
```

Tribute to episode 67, "The Empath"

```
     67

 THEGEM
  MIGHT    where EMPATH is greatest
   HEAL
    THE
  _____

 EMPATH
```

Tribute to episode 68, "Elaan of Troyius"

```
      68
   KIRK
 REACHES    solve in base 16 and maximize TROYIUS
 ELAANOF
 _____

 TROYIUS
```

Tribute to episode 69, "Whom Gods Destroy"

```
        69
      KIRK
     GARTH    solve in base 12 and maximize the great game of CHESS
      CODE
     _____

     CHESS
```

Tribute to episode 70, "Let That Be Your Last Battlefield"

```
        70
      BELE    solve in base 12 and minimize LOKAI
       NOT
      SAME
        AS
     _____

     LOKAI
```

Tribute to episode 71, "The Mark of Gideon"

```
        71
     LOTSOF
     PEOPLE    solve in base 12 and maximize GIDEON
       LIVE
         ON
     _____

     GIDEON
```

Tribute to episode 72, "That Which Survives"

```
        72
     LOSIRA    solve in base 14 where PLANET is maximum
      STILL
     GUARDS
     _____

     PLANET
```

Tribute to episode 73, "The Lights of Zetar"

```
        73
      MIRA   solve in base 12 and maximize LOVE
      LOVE
        OF
     ____

     SCOTT
```

Tribute to episode 74, "Requiem for Methuselah"

```
        74
FLINT       solve in base 14
   AND      since Flint lives so long, we maximize FLINT
  KIRK
  LOVE
_____

 RAYNA
```

Tribute to episode 75, "The Way to Eden"

```
        75
  THE
  WAY       where WAY is longest (greatest)
   TO
_____

 EDEN
```

Tribute to episode 76, "The Cloud Minders"

```
         76
  ELITE
   LIVE     solve in base 12 and maximize CLOUDS
 INCITY
     IN
_____

 CLOUDS
```

Tribute to episode 77, "The Savage Curtain"

```
      77
GOOD  given the significance, we maximize TEST
  VS
EVIL
_____

TEST
```

Tribute to episode 78, "All Our Yesterdays"

```
    78
    IN
  PAST    solve in base 12 and since Spock reverts to the savage Vulcan
  KIRK    of the past, we maximize SPOCK
 SPOCK
_____

 MCCOY
```

Tribute to episode 79, "Turnabout Intruder"

```
     79
   KIRK
   ACTS    since he is only partially a girl, we minimize GIRL
   LIKE
     A
 _____

   GIRL
```

Solutions

First pilot

```
 9   7   4   6  11

    11   0   5

 2   3  10   5

 9   7   8   5

 7   1   3  11
 _____

11   3   4   6   1
```

1.
```
        1
     2749
       52          where 4 and 5 can interchange
     7071
     ─────
     9873
```

2.
```
                    2
12  11   5   0   6   5   1
     3   7  10   4   9   0
    10   6  12  11   6   1
    ─────────────────────
13  11   5   9   8   6   4
```

3.
```
                3
     5  11   2   6   2
     7   4   9   8   7
            11   8   0
             4   7   2
    ─────────────────────
 1   2   3   4   6   2
```

4.

```
      4
   9083
     42
     14
   6593
  _____

  15736
```

5.

```
      5
     84
    742     where 2 & 3 can interchange
    957
     73
  _____

   1861
```

6.

```
      6
   6840     where 0 & 9 can interchange
  17339
  _____

  24185
```

7.

```
                7
           3    4
  15 10   9  13
       9  7    5     where 4 and 6 can interchange
  13 10   5    6
  14  9   1    5
  _____

  2 12  7  11   8
```

8.

```
        8
     7363
      294
        9
     8368
    _____

    16042
```

9.

```
                  9
     11   7   2   3   5
              9   7   0   1
          9   6   4   1  10
              9   7   0   1
     _____

      1  10   8   8   6   2
```

10.

```
                  1   0
     12  10   8   3   7
              4  10   6
          1   0   3   9   9
     _____

     13  11   2  10   8
```

11.

```
                  1   1
          9   4   2   0
          9   6   0   5
          1   8   2   0
                 11   4
     _____

     11   8   7   4  10
```

12.

		1	2
7	8	10	9
1	3	0	10
2	8	4	6
11	8	5	3

where 1 & 2 and 10 & 6 can interchange

13.

```
     13
    703
   5028
     72
   4692
  ------
  10508
```

14.

				1	4
10	4	6	6	0	6

					1	4
10	4	6	6	0	6	
1	8	5	0	10	1	5
2	7	9	7	5	3	4

15.

```
     15
   2610
   4585
  84361
  ------
  91571
```

16.

```
       16
   836819
  7505616
  -------
  8342451
```

17.

					1	7
				1	7	8
	2	3	9	4	0	8
					5	11
	12	5	1	7	5	2
1	0	8	10	13	6	8

18.

```
    18
  9759
   104
  3850
    70
  ─────
 13801
```

19.

							1	9
5	1	7	11	7	10	10	7	9
1	2	4	1	8	11	4	8	0
6	4	0	1	4	10	3	5	6

20.

		2	0
		9	3
8	2	4	11
		6	0
10	9	3	11	11
11	5	7	10	1

where the 4 & 6 can interchange

21.

				2	1	
11	1	5	6	0	8	
			6	2	6	
1	7	10	3	0	7	
1	0	9	4	3	5	10

22.

```
      22
    9806
     410
    1773
      46
    ─────
   12057
```

23. There are two solutions

		2	3
2	4	2	12
4	9	0	7
	13	5	
8	13	10	
6	9	12	6
1	0	5	3	1

		2	3
2	6	2	12
6	7	0	8
	13	5	
10	13	11	
4	7	12	4
1	0	5	3	1

24.

```
      24
     960
    8522
    9652
    2590
   ───────
   21748
```

25.

```
          2   5
      9   0   9
  1   0   8  11  12     where 6 & 8 can interchange
  2   4   6   0  11
                 12
  ─────────────────────
  3   5  10   2   7
```

26.

```
              2   6
              6   1
      3  10   5   0   7
 10   5   5   2   6   9
  6  10  10   9  10   9
                  1   6
  ─────────────────────────
  1   5   8   2   6   4   2
```

27.

```
                  2   7
      4   0   6  10   6     where 3 & 7 can interchange
 13  10   6   4   9  14
 14   8   4   0   6   7
 14   1   6   2  14  14   3
  ─────────────────────────────
 15  13  12  13  10  11   5
```

28.

```
    28
  9109
   845
 42867
   315
 ─────
 53164
```

29.

```
              2   9
    13 11 10  3   9
              6  13
     5  0  6 12   1
             10  12
    ──────────────────
     1  4 12  4   8   2
```

30.

```
        3   0
     9  1   0   9
        8   4   5     where 4 & 6 and the 2 & 3 can interchange
  4  2  8   6   7
        3   1   5
  ──────────────────
  5  3  1   6   4
```

31.

```
              3   1
     3  9  6  4  13   7
              3   0  11   0     where 1 & 7 and 2 & 4 can interchange
    10  3  0  2   9   1
    ──────────────────────
    13 12  9  8   8   9
```

32.

```
    32
    64
  9049
    63
 ─────
 18257
```

33.

```
      33
    3948
    9870
  358868
  358868
  _____

  731587
```

34.

```
    34
  7471
  7261      where 6 & 9 can interchange
    93
  5228
  _____

  20087
```

35.

```
        3   5
      7 0   8
    6 10 5  6     where 3 & 7 and 1 & 2 can interchange
    1  2 8  4
       3 10 9
    2 10 5  8
    _____

  11 10 10  4
```

36.

```
                  3   6
    13 12 11  9   8   0
        5  4  6   4  10
        3 11  0  12   0
    _____

  1  0  7 13  3   0   2
```

37.

```
    37
   192        where 1 & 3 can interchange
  7386
   929
  ─────
  8544
```

38.

```
            3    8
       1    3    8   12    7      where the 2 & 3 can interchange
  12   7    5    8    5    4
   1   4    2    6    7    9
  ─────────────────────────
  13  12   11   10    1    0
```

39.

```
                 3    9
       11  10    8    5    3      where the 2 & 3 can interchange
                10    9    2
  10    6  10    9    2   10
  ───────────────────────────
   1   10   8    5    9    0   11
```

40.

```
             4    0
         3  11    5       where 1 & 3 & 5 can interchange
      2  11  11    3
  10   9   7    0    6
             8    1
  ──────────────────
  11    0  11   11    3
```

41.

```
      4  1
   8  3  0  6
      8  4  6
      6  1  0
      7  2
   ─────────
   9  8  7  5
```

42.

```
           42
          899
        59660
        78472
          899
        _____

       139972
```

43.

```
        43
    7658
        42      where 0 & 2 and 1 & 4 can interchange
      910
      ____

    8653
```

44.

```
                    4    4
            11   1   4   11
                10   9    3
         7   5   4   4    8
            10   2  11    7
    10   9   3  10   3    3
    _____

    11   6   7   6   2    0
```

45.

```
                    4    5
             8   5   2    8
         6   9   6   2    7
    10   9   0   8   1    3
             7   2    0
    _____

    11   4   7   3   0   11
```

42

46.

```
        46
         1
     54626
     73806
    _____

    128479
```

47.

```
                    4    7
                   11    8
     10   8  10    5    0    2        where the 4 & 7 can interchange
          7   5   11    1    8
          4  11    5    0   10
         _____

     11   9   3   10    6   11
```

48.

```
                         4    8
      1  11   2    6    6   11    7
      3   1   0    2    6    2    4
     _____

      5   0   2    9    1    6    7
```

49.

```
                    4    9
                         9
          8   6   10    0
      9  11   3    5    2    3        where 4 & 8 can interchange
          9   2    6    0   11
          4    6   11    3
     _____

     10   9   7    1    5   11
```

50.

```
        50
        52
      4857
      3452
      1066
     _____

      9477
```

43

51.

```
              5   1
       3  11 13   3
       2   5  8   1
   12 11  10  9   1
   ─────────────────
   13  4   0   7   6
```

52.

```
         52
        145
       1447
      28635
     ───────
      30279
```

53.

```
                    5   3
     10   5   6   5   9   6     where the 3 & 9 can interchange
      5   7   0   8   3   2
     ─────────────────────────
   1  4   0   7   2   5  11
```

54.

```
        54
      8794
        30
      7254
     ──────
     16132
```

55.

```
               5   5
        5  10  5  10
            9   0   6     where the 4 & 6 can interchange
        3   9   8   4
   10  11   2  11   0
   ──────────────────
   11   9   8   7   1
```

56.

```
                    5   6
                   10   9
 5   6   1   7  10   7  10
         4   0   9   8   3
                   10   9
 5   6   1   7  10   7  10
_____
11   0   7   4   9   2  11
```

57.

```
                 5   7
        10   6   7  10
         9   3   2   4
12   5   4   4   8
         7   8   5
_____
13  11   8   0  10
```

58.

```
        58
      4564
     21487
     19508
        17
    _____
     45634
```

59.

```
        59
     54302
     96047
      9730
    _____
    160138
```

60.

					6	0
	8	9	10	10	9	11
	4	5	13	0	2	11
		11	1	9	7	8
13	12	11	6	12	2	

61.

				6	1
	12	13	6	11	
	12	13	6	11	
	12	13	6	11	
			1	5	9
	13	2	0	7	
3	10	2	4	8	

where 2 & 5 and 4 & 6 can interchange

62.

```
  62
 185
  23        where E can also be 0
 976
─────
1246
```

63.

```
    63
690389
   964
 25517
    16
──────
716949
```

64.

			6	4	
	9	10	7	9	
10	8	5	1	2	
			5	0	
11	6	4	8	3	

where the R & U and the T & F can interchange

46

65.

```
     65
   4574
  16171
   4500
    398
  ─────
  25708
```

66.

```
     66
   6906
     14
  73552
     14
  ─────
  80552
```

67.

```
      67
  283931
   17982
    8365
     283
  ───────
  310628
```

68.

```
                6   8
          2   9  14   2
  14   1   5   4  11   1   8     where 3 and 11 can interchange
   1  13   5   5   3  10   6
  ─────────────────────────
  15  14  10  12   9   0   8
```

69.

```
           6   9
       4   3   1   4     where 2 & 3 and 0 & 8 can interchange
  10   6   1   8   9
      11   2   0   7
  ──────────────────
  11   9   7   5   5
```

70.

```
              7   0
     6    7   1   7
         10   0   5
     4    8   3   7
              8   4
    ──────────────────
  1  0    2   8  11
```

71.

```
                  7   1
    9   1   4   6  1   7
    2   8   1   2  9   8
            9  10  5   8
                   1   0
   ──────────────────────
  11 10   3   8   1   0
```

72.

```
                      7   2
   12   0   8   4   6 11
        8   5   4  12 12      where 0 & 3 can interchange
    1   3  11   6  10  8
   ──────────────────────
  13  12  11   2   9   5
```

73.

```
              7   3
     4  11   0   2      where the 2 & 6 can interchange
    10   7   5   9
              7   6
    ──────────────
   1   3   7   8   8
```

74.

```
               7   4
   12  11   4   7   8
           10   7   5      where the 5 and 6 can interchange
        1   4  12   1
       11   3   0   6
   ──────────────────
  13  10   9   7  10
```

48

75.

```
     75
    321
    986
     35
   ──────
   1417
```

76.

```
              7   6
      9   6  10   3   9
          6  10   2   9
 10   7  11  10   3   5
             10   7
 ───────────────────────
 11   6   1   8   4   0
```

77.

```
     77
   1445      where 0 & 5 and 3 & 4 can interchange.
     37      Note that the 3 & 4 is a double interchange
   8320
   ──────
   9879
```

78.

```
           7   8
           7   3
    8   0  10   4     where 3 & 4 can interchange
    1   7   2   1
 10   8   9   6   1
 ──────────────────
 11   6   6   9   5
```

79.

```
   79
 2042      where 5 & 6 can interchange
 1895
 3026
    1
 _____

 7043
```

Explanations of the cartoons

The first is simply the view screen of the Enterprise and contains the catch phrase of the original series in alphametic form.

In the mathematical construct known as Boolean algebra, the two expressions seen in the second cartoon are known as the "absorption laws." Since the computer-based entity known as Landru absorbs people into "the body" these laws are a fit into the world controlled by Landru.

The function on the left of the expressions is called the Riemann Zeta function and in this case it is set equal to R. This is a reference to episode 73, "The Lights of Zetar."

ADDITIONAL BOOKS PUBLISHED BY CHARLES ASHBACHER

Assistant editors

Rachel Pollari

Jennifer Corrigan

Artwork

Caytie Ribble

Jenna Richardson

TOPICS IN RECREATIONAL MATHEMATICS 1/2015

ISBN 978-1507603215

TOPICS IN RECREATIONAL MATHEMATICS 2/2015

ISBN 978-1508617099

TOPICS IN RECREATIONAL MATHEMATICS 3/2015

ISBN 978-1511641005

TEN YEAR CUMULATIVE INDEX to the JOURNAL OF RECREATIONAL MATHEMATICS

ISBN 978-1508936800

ALPHAMETICS AS EXPRESSED IN RECREATIONAL MATHEMATICS MAGAZINE

ISBN 978-1508538134